DISCOVER!

EYE ON ECOSYSTEMS

DO NATURAL DISASTERS CHANGE ECOSYSTEMS?

THERESA EMMINIZER

Enslow PUBLISHING

Please visit our website, www.enslow.com.
For a free color catalog of all our high-quality books, call toll free 1-800-398-2504 or fax 1-877-980-4454.

Cataloging-in-Publication Data
Names: Emminizer, Theresa.
Title: Do natural disasters change ecosystems? / Theresa Emminizer.
Description: New York : Enslow Publishing, 2023. | Series: Eye on ecosystems | Includes glossary and index.
Identifiers: ISBN 9781978532755 (pbk.) | ISBN 9781978532779 (library bound) | ISBN 9781978532786 (ebook)
Subjects: LCSH: Natural disasters–Juvenile literature. | Biotic communities–Juvenile literature. | Biodiversity–Juvenile literature.
Classification: LCC GB5019.E46 2023 | DDC 363.34'92–dc23

Published in 2023 by
Enslow Publishing
2544 Clinton Street
Buffalo, NY 14224

Designer: Tanya Dellaccio
Editor: Theresa Emminizer

Photo credits: Cover A_Lesik/shutterstock.com; p. 5 yerv/Shutterstock.com; p. 7 ImageBank4u/Shutterstock.com; p. 9 FotoKina/Shutterstock.com; p. 11 Kirill Skorobogatko/Shutterstock.com; p. 12 Eric Lysenko/Shutterstock.com; p. 13 AP Images; p. 15 Minerva Studio/Shutterstock.com; p. 17 Andriy Solovyov/Shutterstock.com; p. 19 kavram/Shutterstock.com; p. 21 Matt Valentine/Shutterstock.com.

Printed in the United States of America

CPSIA compliance information: Batch #CWENS23: For further information contact Enslow Publishing at 1-800-398-2504.

CONTENTS

Boldface words appear in Words to Know.

EARTH'S ECOSYSTEMS

An ecosystem is a natural community that includes living and nonliving things. There are many different ecosystems on our planet. Natural disasters are events in nature, such as floods or wildfires, that cause harm. Natural disasters can wipe out ecosystems or change them forever.

NATURAL DISASTERS HAPPEN VERY SUDDENLY.

NATURAL DISASTERS

About 400 natural disasters happen across the planet each year. Sometimes natural disasters are caused by Earth's natural **processes**. The movements of the moon, Earth, and tides can lead to natural disasters. However, natural disasters are caused by human activities too.

VOLCANIC ERUPTIONS ARE CAUSED BY EARTH'S NATURAL PROCESSES.

CLIMATE CHANGE

Human activities are causing changes to Earth's climate, or average weather **conditions** over a period of time. This is known as climate change. Climate change increases the number of natural disasters. It can also make natural disasters more intense, or strong, when they happen.

HIGHER HEAT AND MORE RAINFALL HAVE CREATED STRONGER STORMS.

GEOLOGICAL DISASTERS

Geological disasters are land-based natural disasters such as earthquakes and avalanches. An earthquake is a shaking of Earth's surface. It's caused by movements in Earth's **crust**. An avalanche happens when a large amount of ice, snow, and rocks fall quickly down a hill or mountain.

A LARGE AVALANCHE CAN MOVE FASTER THAN 200 MILES (321.9 KM) PER HOUR!

HYDROLOGICAL DISASTERS

A hydrological natural disaster is one that has to do with water, like a flood or tsunami. A tsunami is a huge, fast-moving sea wave. It rises high and comes crashing onto shore, ripping up trees and washing away entire beaches.

A TSUNAMI CAN MOVE AS FAST AS 500 MILES (804.7 KM) PER HOUR!

METEOROLOGICAL DISASTERS

Meteorological disasters have to do with the weather or climate. A tornado is a **violent** windstorm that moves over the land. Tornadoes can destroy ecosystems, ripping up trees, digging into the earth, and killing animals as they go.

TORNADO WINDS MAY REACH 300 MILES (482.8 KM) PER HOUR.

WILDFIRES

Climate change has led to longer droughts, or dry periods. These dry conditions make wildfires more likely. Wildfires can be started by lightning, but they're often started by people. They can spread quickly, destroying **habitats**, killing plants and animals, and **releasing** harmful gases.

SOME WILDFIRES HELP ECOSYSTEMS BY CLEARING AWAY HARMFUL THINGS.

LASTING HARM

Some ecosystems can rebound, or bounce back, from natural disasters. Some are destroyed forever. Some animals are pushed closer to extinction, or dying out completely. During 2019 and 2020, wildfires in Australia lead to the death of about 1 billion animals.

ONE-THIRD OF THE WORLD'S KOALA POPULATION WAS LOST TO THE AUSTRALIAN WILDFIRES.

STAYING SAFE

Natural disasters can be scary. But being prepared helps us stay safe! **Scientists** track Earth's conditions to try to learn when natural disasters may be coming. If a natural disaster does happen, there are **emergency** plans to help keep people safe.

WE CAN STAY SAFE FROM NATURAL DISASTERS.

WORDS TO KNOW

condition: The state of something.

crust: The outer shell of a planet.

emergency: An unexpected situation that needs quick action.

habitat: The natural place where an animal or plant lives.

process: A series of steps or actions.

release: To let something out.

scientist: A person who studies the way things work and the way things are.

violent: Having to do with the use of force to cause harm.

volcanic eruption: The bursting forth of hot, liquid rock from within Earth.

FOR MORE INFORMATION

BOOKS

Harmon, Daniel E. *Tsunamis.* New York, NY: Britannica, 2019.

Johanson, Paula. *Earthquakes.* New York, NY: Britannica, 2019.

WEBSITES

Ready
www.ready.gov/kids/be-ready-kids
Find out more about natural disasters and how you can be prepared!

Wiz Kids
www.weatherwizkids.com
Learn more about what causes natural disasters.

Publisher's note to educators and parents: Our editors have carefully reviewed these websites to ensure that they are suitable for students. Many websites change frequently, however, and we cannot guarantee that a site's future contents will continue to meet our high standards of quality and educational value. Be advised that students should be closely supervised whenever they access the internet.

INDEX